The Structures of Crystals

Student Monographs in Physics

Series Editor: Professor Douglas F Brewer
Professor of Experimental Physics, University of Sussex

Other books in the series:

Microcomputers
D G C Jones

Maxwell's Equations and their Applications
E G Thomas and A J Meadows

Oscillations and Waves
R Buckley

Fourier Transforms in Physics
D C Champeney

Kinetic Theory
J M Pendlebury

Diffraction
C A Taylor

Vectors and Vector Operators
P G Dawber

The Structures of Crystals

A M Glazer

*Lecturer in Physics at the Clarendon Laboratory, Oxford
and
Fellow of Jesus College, Oxford*

Adam Hilger, Bristol

026 39099

PHYSICS

British Library Cataloguing in Publishing Data

Glazer, A.M.
  The structures of crystals.
  1. Crystals
  I. Title
  548'.81   QD921

  ISBN 0-85274-825-6

Published under the Adam Hilger imprint by IOP Publishing Limited
Techno House, Redcliffe Way, Bristol BS1 6NX, England

Typeset by KEYTEC, Bridport, Dorset
Printed in England by WBC Print Limited

Contents

Preface

When I was approached to write this small book about crystals, I thought it was going to be straightforward. However, once I started, it became painfully clear that I was in trouble. Torn between wanting to define everything in painstaking detail and, at the same time, to set out everything in the simplest and clearest possible way, I had to make some ruthless decisions. This book is intended for physics undergraduates, typically in their first or second year of study, and because of space limitations, I could not apply the full rigour of the subject of crystals and symmetry, even though I personally find it very beautiful and satisfying. I therefore have had to jettison many favourite proofs, explanations and definitions in favour of an informal style, which I hope will make the text easy to read. In many places I have used a lot of 'hand-waving' to explain ideas that could be more formally treated by mathematics. In several cases, I give formulae without proof, so that we can concentrate better on their consequences: in any case, formal proofs can always be found in the many standard texts on crystallography. There are four chapters in this book. The basic ideas of symmetry are first tackled with an introduction to the classification of crystals into the seven systems. This is followed by a discussion of lattices, which I have deliberately separated from actual crystals in order to make the distinction clear. The third chapter introduces the crystal structure concept, and illustrates this with descriptions of some simple structures. Finally, the diffraction of x-rays by crystals is discussed briefly. We cannot do justice here to this important topic, but the material presented in the last chapter should be helpful in giving the reader extra insight when reading other textbooks on x-ray diffraction. I have tried to cover particular concepts that are often badly treated in undergraduate texts. The most common confusion I have found, for that matter even among working physicists, is in the distinction between lattice points and atoms, and I hope that this book will make the distinction clear. Throughout, I have highlighted specific problems, which you should tackle before reading on. They are all elementary, and in most cases I give the answers right away, as they are intended more to emphasise points that I think are important.

I admit unashamedly that my approach to the teaching of crystallography is slightly unconventional. I make liberal use of the idea of convolution and Fourier transforms to define a crystal structure and to explain the diffraction patterns produced when x-rays, neutrons or electrons are scattered by a crystal. Although convolution is excluded from many undergraduate optics courses, I nevertheless feel that it is inherently easy, and even if used visually, rather than mathematically, can pay enormous dividends in the basic understanding of crystals and diffraction in general. I hope that you will agree with me by the time you get to the end of this book. Finally, I must thank my daughter, Alison, for her drawings of the chicken and sharks in chapter 3, my son Richard for proof reading, Pam Thomas for the x-ray photographs in chapter 4, and Dr N Swindells and Dr K Stadnicka for their comments on the manuscript.

Crystal Systems 1

1.1 Introduction

If you visit a museum that has a collection of mineral specimens, you cannot fail to be struck by the vast range of colours, lustres and regular shapes of the crystals on display. I still remember my first visit, at the age of seven years, to the Geological Museum in London, and the impression created on me has stuck to the present day. Crystals have always been recognised as being distinct from other forms of matter for as far back as history has been recorded. The word 'crystal' is derived probably from the Greek word meaning 'ice'. We are all able to recognise crystals when we see them, such as in the salt or sugar at the dinner table, but it is less widely realised that most solid material can occur in the form of crystals, even if very tiny. Crystals are found throughout the natural world as the stable forms of metals, organic and inorganic chemicals, natural and synthetic products and in biological organisms. Even bone has something of a crystalline content to it.

These days, the synthetic growth of crystals is an important part of the modern world. Truly stupendous sums have been spent on the growth of pure and perfect crystals of silicon, upon which most of the present boom in information technology depends. The list of new and interesting technologically important crystals grows ever longer. We can expect to see increasing use made of new crystals in computers, communications and in other high-technology areas. GaAs crystals will soon begin to replace silicon in the semiconductor industries. Hundreds of pages of information can now be stored holographically within a single crystal, and in communications, crystals are currently being considered for their optical changes when sound waves are passed through them. The list of the uses of crystals is endless.

It has been known for some time that crystals have regular, plane faces (as illustrated, for example, by borax and calcite in figure 1.1) with fixed angles between them. Furthermore, the angles between the faces of a crystal of, say, borax are the same no matter from which part of the world it comes. The faces themselves may be of different sizes, depending on the conditions under which the crystals have grown, but the relationships between them remain fixed.

1

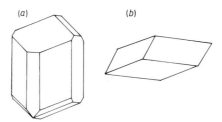

Figure 1.1   A crystal of (*a*) borax
($Na_2B_4O_7 \cdot 10H_2O$) and (*b*) calcite ($CaCO_3$).

This relationship between faces led to early theories about how crystals were constructed. Huygens, in his Treatise on Optics (1690), thought of crystals of the mineral calcite as being built from small ellipsoidal units of some kind, and with this picture he was able to explain how it was that calcite crystals always cleaved into rhombs with the resulting faces still preserving the same relationships. In 1784, Haiiy published his 'Essai d'une Théorie sur la Structure des Cristaux', and thus began the science of crystallography. He developed Huygen's concept to explain the relationships between the faces. For the next hundred years, an enormous amount of work was carried out identifying, cataloguing and generally trying to understand crystals. For the most part, the techniques used in this study were confined to chemical analyses and optical observations with microscopes and goniometers to measure interfacial angles accurately. At the same time, it was noticed that many physical properties seemed to be tied in some way to the observed shapes (**morphology**) of crystals.

An outstanding example of this is given by the phenomenon known as **optical activity**. This was discovered in 1811 by Arago, who noticed that sunlight reflected from a plate of quartz and then passed through another crystal of quartz, appeared to show colour changes when the second crystal was rotated. Biot, soon after, showed that the plane of polarisation of light in such crystals underwent a rotation, so it was rotated either to the right (dextrorotation) or to the left (laevorotation). The same effect was found also in solutions of many organic substances. Pasteur, during his research into certain problems afflicting the French wine industry, examined crystals of sodium ammonium tartrate under the microscope; these had been derived from the tartaric acid found on the insides of wine vats. Using a microscope, he was able to separate the crystals into two forms (figure 1.2(*a*)), each one the mirror image of the other. On dissolving in water, each type of crystal gave solutions which were either dextro or laevorotatory. From this observation, Pasteur made the important leap of imagination, namely that the difference in optical rotation must be due to some asymmetry in the tartaric acid molecules, and that the external shapes of crystals therefore must be telling us something about the way in which atoms and molecules are arranged within the crystal. A similar link between morphology and the sense of optical rotation was also discovered earlier by the English astronomer, Herschel, in crystals of quartz (figure 1.2(*b*)).

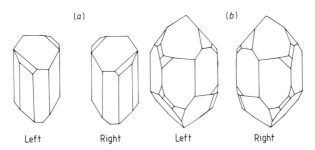

Figure 1.2 Left- and right-hand crystals: (*a*), sodium ammonium tartrate; (*b*), quartz.

During the nineteenth century, many famous scientists were fascinated by crystals, and experimentation and theoretical development was intensively pursued. However, they were hampered by the fact that, at that time, there was no way to 'see' the internal structures of crystals, nor, incidentally, was it by any means sure that atoms or molecules had any physical existence. Even the nature of light as a wave- or particle-like phenomenon was in question. The situation changed irreversibly in 1912 with the discovery of x-ray diffraction by Max von Laue working in Munich. This marked the start of the modern science of crystallography, since it led to techniques that now enable the atomic arrangement of a crystal to be worked out in its entirety.

1.2 Symmetry

The most obvious characteristic of crystals is that of **symmetry.** This is not an easy concept to define, although most of us seem to have an innate sense of appreciation of symmetry. It has been said that 'symmetry is death', and if you think about it, you will soon understand how true this is. If something has symmetry, it is therefore static, unchanging, frozen as in death. Indeed, it can be argued that it is change in symmetry or the so called breaking of symmetry that leads to new discoveries, to development, to vitality and to progress. For instance, music that repeats continuously is very symmetric, but, for most of us, it is boring, whereas when music changes dramatically, our interest is aroused.

There is a curious thing about symmetry: it shows discrete behaviour. Consider a ball bouncing on a table. Before it hits the table it is spherical. However, the moment it touches the table, it becomes slightly squashed and its spherical symmetry is instantaneously broken. The ball keeps its 'squashed' symmetry until it has bounced off the table and has regained its spherical shape. Thus, at any instant, the ball either has spherical symmetry or it has a 'squashed' symmetry: it cannot be halfway between. Such a sudden change in symmetry accompanies any change of state, and so a study of the breaking of symmetry is of particular importance in understanding dynamic processes such as phase transitions. However, although the really interesting processes involve symmetry breaking, we first have to understand the nature of symmetry itself, especially if we are to learn how to describe the structures of crystals.

To deal with symmetry in a more scientific way, consider the hexagonal arrangement of spheres shown in figure 1.3. Suppose that you were to look away from this book for a moment, and that somehow I could then contrive to rotate this figure through 60° about the axis OZ through its centre. On looking back at the page, you would not be aware that anything had changed, even though *I* know that a change was made. We could repeat the game over and over again, always with the same result. Note, however, that if the process of rotation by 60° is carried out an integral multiple of six times, the object really *is* back in its original position (but only *I* know that!).

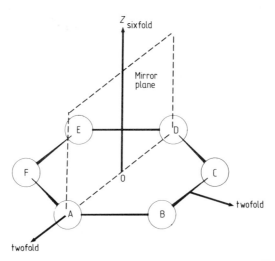

Figure 1.3   A hexagonal ring showing many symmetry elements.

This type of process is called a **symmetry operation**. Each symmetry operation alters the crystal (or any object for that matter), and puts it into a **symmetry-related position**. The above symmetry operation is a sixfold **rotation** about the point at the centre of the hexagonal ring. Note that it is conventional to consider all rotation operations as taking place in an anticlockwise sense. Symmetry operations that operate about a fixed point in space are known as **point symmetry operations**. We can also say that the hexagonal ring has a sixfold axis of rotation perpendicular to its plane and passing through its centre. The axis is referred to as a **symmetry element**. Many scientists tend to mix up the concepts of 'operation' and 'element', although strictly speaking they are different ideas. In our example, there is one sixfold symmetry element (the axis of rotation), but six sixfold operations (six successive rotations of $60°$ to bring the object back to its initial state).

There are two main systems of notation currently used to describe symmetry operations, one called the Schoenflies notation and the other the International notation. Historically, the Schoenflies system has been used principally by chemists and physicists, and the International system by crystallographers. It is true to say that the International notation is gaining ground as more scientists learn to appreciate its conciseness. A full description of the symbols lies outside the scope of this book, and is, in any case, well covered in several standard texts. The International notation is explained rigorously in the crystallographer's bible: International Tables for Crystallography. In this book, I shall just give you a 'flavour' of what the International notation is like. The notation for the six successive sixfold operations in the hexagon is quite simply 6, 6^2, 6^3, 6^4, 6^5 and 1. Thus 6^2 is equivalent to the sixfold operation applied twice in succession, i.e. rotation through $120°$. Why 1? Well, applying the operation six times i.e. 6^6, brings us back to the start, and so it is equivalent to not having done anything at all. That is, it is a onefold or **identity** operation. Other operations can be found for the hexagonal ring. For instance, the 6^3 operation is the same as rotating through $180°$, i.e. it is the same as a twofold operation, symbolised simply by 2. The 6^2 operation is equivalent to a 3 operation, and the 6^4 to a 3^2.

Problem 1.1   *What is the relationship between the 6 and 6^5 operations?*

We normally take the rotation operations to act counterclockwise to our line of sight. The 6^5 operation is the same as a single sixfold operation carried out in the unconventional clockwise sense, i.e. the mathematical inverse. There are also twofold operations that turn the hexagon over about an axis lying in the plane and joining opposite vertices, such as A and D in figure 1.3.

Problem 1.2   *How many of these new twofold operations are there? And are there any other types of twofold operations?*

The sixfold rotation acts not only on the hexagonal ring, but also on all the other symmetry elements present. The result of this is to produce three twofold axes in the plane of the ring at 120° to one another. These are along the lines AD, BE and CF. In addition, there are three other twofold axes running perpendicular to the edges of the hexagonal ring and passing through the centre.

There are two other operations that I wish to draw to your attention. First, the hexagon can be reflected across a plane running perpendicular to the hexagon and through opposite vertices, say A and D. Thus one side of the hexagon can be related by reflection to the other. For this reason, the operation is called a **mirror operation**, and is given the symbol m. An important aspect of mirror operations is that two objects related by a mirror cannot be turned into each other by rotations. Thus, there is no way to turn your left hand to look the same way as your right hand just by rotating one hand. Try it! Look also at the two crystals of sodium ammonium tartrate and quartz in figure 1.2. Can the right-hand forms be turned into the left-hand forms by rotation? Think also about a nut and bolt. Suppose that the screw thread on each is of the same hand. As you know from experience, it does not matter which way the nut is threaded on to the bolt; it still fits. However, if you have a bolt with a left-hand thread and a nut with a right-hand thread (i.e. one thread is the mirror image of the other), there is no way that you can screw the bolt into the nut.

Problem 1.3 *When you look at yourself in a mirror, it appears that the right side of your body becomes the left and your left becomes the right in the image. However, your head and feet are not interchanged. Does this suggest that a mirror somehow only works in the horizontal direction and not in the vertical?*

The other type of operation is called **centrosymmetry**, and is denoted by the symbol $\bar{1}$. This is best described mathematically by the mapping of coordinates $(x, y, z) \rightarrow (-x, -y, -z)$. Hold your left hand out in front of you, palm down, and your right-hand palm up and with the fingers pointing towards you (figure 1.4). You can see that the tip of your right thumb is related to the tip of your left thumb through a line passing through a point midway between your hands. The same relationship is true for your fingers, and indeed for all parts of your hands (ignore the blemishes!). The pair of hands forms a centrosymmetric arrangement, and the point midway between them is called the **centre of symmetry** or **centre of inversion**. A single hand is not centrosymmetric. Notice that, as with mirror operations, a centre of symmetry turns a right-hand object into a left-hand one, and vice versa. Both are examples of inversion operations.

Crystals of quartz and sodium ammonium tartrate are not centrosymmetric. In fact, it can be shown that the occurrence of optical activity cannot arise if the crystals are centrosymmetric, once again pointing to the close link between physical properties and crystal symmetry.

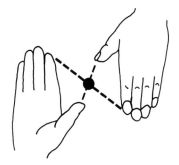

Figure 1.4 A centre of symmetry.

Finally, you should be aware that crystallographers use another type of operation known as **rotation–inversion**. We shall not deal with it in this book, except to say that it is denoted by a bar placed over the rotation symbol. We have already seen one of these, $\bar{1}$. The mirror operation **m** is equivalent to $\bar{2}$ and so it too is a point operation.

1.3 Crystal Systems

The early crystallographers noticed that crystals of different substances often had similar shapes, and this led to a simple classification of all crystals into the so called **seven crystal systems**. These are listed in table 1.1. Figure 1.5 shows the right-hand screw convention for the labelling of axes and interaxial angles. Make sure that you learn this convention.

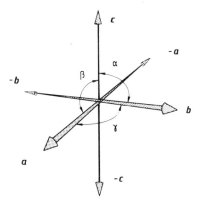

Figure 1.5 The system of labelling of axes and angles (from Burns and Glazer).

Table 1.1 The seven crystal systems.

Name	Symmetry elements	Axial restrictions
Triclinic	Identity or centre of symmetry	—
Monoclinic	One twofold axis or one mirror or a twofold perpendicular to a mirror	$\alpha = \beta = 90°$
Orthorhombic	Three twofold, or three mirrors, or two mirrors plus twofold	$\alpha = \beta = \gamma = 90°$
Tetragonal	One fourfold axis (or $\bar{4}$)	$a = b$ $\alpha = \beta = \gamma = 90°$
Cubic	Four threefold axes	$a = b = c$ $\alpha = \beta = \gamma = 90°$
Trigonal	One threefold axis	$a = b$ $\alpha = \beta = 90°$ $\gamma = 120°$
Hexagonal	One sixfold axis	$a = b$ $\alpha = \beta = 90°$ $\gamma = 120°$

First of all, there are many different materials that crystallise in the form of cubes, e.g. iron pyrites and fluorite (figure 1.6(*a*)). Thus, despite the fact that they are chemically totally different substances, it seems natural from the crystallographic point of view to put them into the same classification system. You will not be surprised to learn that it is called the **cubic system**. You will notice from this figure that we can set up three mutually perpendicular axes *a*, *b* and *c* running through the centres of each face of the cube, and the lengths of these axes (given by the distance from the centre of the crystal to its faces) are equal. However, we now encounter a problem: other crystal shapes may turn up with the same chemicals. Thus iron pyrites and fluorite often are found in the form of octahedra (figure 1.6(*b*)). Iron pyrites is also found in the form of so called pyritohedra (figure 1.6(*c*)). Obviously, it would not be helpful to put all these different crystal forms of the same substances into different classifications. Furthermore, there is something common to all these shapes. We can mark three mutually perpendicular equidimensional axes through them, just as with the cube. Look at the symmetry elements in these crystals. There are three fourfold axes along *a*, *b* and *c*, six twofold axes halfway between *a* and *b*, *b* and *c*, and *c* and *a* (parallel to the face

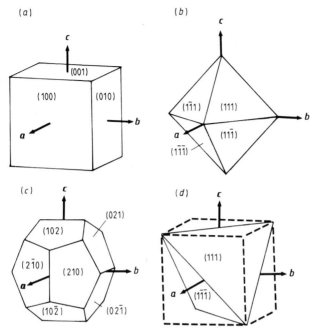

Figure 1.6 Some examples of different crystals in the cubic system: (a), a cube; (b), an octahedron; (c), a pyritohedron; (d), a tetrahedron (and its construction out of a cube).

diagonals), and four threefold axes at equal angles to **a, b** and **c** (along the body diagonals). It turns out that the most important common elements are the four threefold axes, and these define the cubic crystal system. It is possible to have cubic crystals that have no fourfold axes, but they must have the four threefold axes.

Problem 1.4 *Why does a tetrahedron belong in the cubic system?*

A tetrahedron has four threefold axes, each one coming out through the centre of each of its four faces (the faces have triangular shapes for that reason). Also, you can construct a tetrahedron out of a cube by joining up four corners of a cube as in figure 1.6(d). This shows nicely the orientation of the tetrahedron with respect to the cube and thus to the cube axes, **a, b** and **c**.

I must stress the fact that the cubic system is defined in terms of its symmetry, and not in terms of the lengths and angles of the axes. This is because it is quite possible to have a crystal in which $a = b = c$ and the angles $\alpha = \beta = \gamma = 90°$, within the limits of any measurement that we can make, and that accidentally belongs to any crystal system except trigonal or hexagonal. This may become apparent only when we study its physical properties or its arrangement of atoms, i.e. its crystal structure, for this too must obey the symmetry constraint of having four threefold axes. Thus, it is the symmetry that imposes restrictions on the axial lengths and angles, and not the other way about. For this reason, in table 1.1, the seven crystal systems are defined first through their symmetries, with the axial conditions following as a consequence. Many textbooks fail to stress this and, in so doing, incorrectly define the crystal systems.

The cubic system is the most important one as far as physics undergraduates are concerned, but you should at least be aware of the existence of the other six. Let me very briefly run through them.

In the **tetragonal system**, there is a single fourfold axis, thus forcing the *a*- and *b*-axes to be equal in length, with the *c*-axis free to be any length (remember, it could *accidentally* be the same length as *a* and *b*), and with all axes mutually perpendicular. A cube stretched along one of its fourfold axes will lose its threefold symmetry and retain only the fourfold axis, thus becoming tetragonal. An object with three mutually perpendicular twofold axes or mirror planes (or combinations of the two) belongs to the **orthorhombic system**, and as a consequence, the axes are mutually perpendicular but not necessarily of the same length. A brick can be thought of as an example of an object belonging to the orthorhombic system, although it is not actually a crystal.

If there is a single sixfold axis, as in the hexagonal ring used earlier, the crystal belongs to the **hexagonal system**. This now forces the angle between the *a*- and *b*-axes to be at 120° to each other and perpendicular to *c*. At the same time, the lengths of *a* and *b* must be the same. The same constraint is imposed on the axes of **trigonal** crystals, which have a single threefold axis (you should not be surprised about this, since you will recall that a threefold axis is actually contained within a sixfold axis: $6^2 \equiv 3$). Quartz crystals belong in the trigonal system.

If there is only a single twofold rotation or a mirror plane present, the crystal is then **monoclinic**. Now the axes can be any length, the only constraint being that two of the interaxial angles must be 90°.

Finally, objects with no symmetry elements at all, apart from the identity, or with just a centre of symmetry, belong to the **triclinic system**. There are no restrictions then on the axes and angles.

It is a good idea to practice classifying different objects in terms of crystal systems. Try it with anything in the room, and you should find it possible to classify all objects in one of the seven crystal systems.

1.4 Point Groups

Consider what limits the number of symmetry operations for any object. In the hexagonal ring, we saw just a few of the total number of symmetry operations. Now because each operation acts on every other operation, and repeated application of an operation a finite number of times brings us back to the start, there will be a finite number of operations for a hexagonal ring. Actually, there are 24 in all, although we shall not concern ourselves with identifying them all here. The 24 operations can be shown to form a group in the mathematical sense, and because they are all point operations, they constitute what is termed a **point group**. Briefly, the characteristics that make a mathematical group are:

(i) There is an identity element (obviously always present for any object).

(ii) To each operation there corresponds an inverse operation that is also contained within the group. (Remember, 6 is like 6^5, only the opposite way round thus $6^{-1} \equiv 6^5$.)

(iii) The product of any two operations is equal to another operation in the group (e.g. 6 operating on 6 is the same as $6^2 \equiv 3$ etc.).

(iv) The law of associativity is true. That is $(6^2 6^3)6$ is the same as $6^2(6^3 6)$. The point group of a hexagon is called in the International notation 6/mmm (and in Schoenflies notation D_{6h}). A quartz crystal belongs to the point group 32 in the International notation (six symmetry operations in the group, consisting of the identity, 3, 3^2, and three twofold axes perpendicular to the threefold axis). Either one of your hands belongs to point group 1 (there is only one symmetry operation, the identity operation 1), whereas together, as in figure 1.4, they are in the point group $\bar{1}$ (with operations 1 and $\bar{1}$). Your face more-or-less belongs to the point group m (two operations, identity plus the mirror relating both sides of your face). It turns out that in crystals there are only 32 distinct point groups possible in three dimensions, and these are sometimes therefore called the **32 crystal classes**.

1.5 Indexing of Faces

An important part of the classification of crystals is to have some means of identifying the various faces seen on a crystal. The usual way of doing this is to label each face by its **Miller indices**. To define them, consider a plane to cut the three axes *a, b* and *c* (not necessarily orthogonal to one another) at A, B and C (figure 1.7). Let us now define the intercepts OA, OB and OC as fractions a/h, b/k and c/l of the lengths of the axes. The index for this plane is then given as (hkl). Note the convention: the indices of faces or planes are always quoted in round brackets.

Let us take some examples to make this clear. The cube in figure 1.6(*a*) has six equivalent faces. The one at the front cuts the *a*-axis to make an intercept

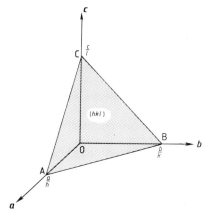

Figure 1.7   The indexing of crystal faces and planes.

at a/h and is parallel to the **b**- and to **c**-axes i.e. it cuts **b** and **c** at infinity. Thus the intercepts are given by:

$$\frac{a}{h} \qquad \frac{b}{0} = \infty \qquad \frac{c}{0} = \infty$$

and the index could then be written as $(h00)$. Conventionally, the smallest possible values of the indices are taken to label the faces of the crystals, and so this cube face is labelled (100). Similarly, the faces perpendicular to **b** and to **c** in the cube are labelled (010) and (001). The remaining three faces make negative intercepts on the axes and so are labelled

$$(\bar{1}00) \qquad (0\bar{1}0) \qquad (00\bar{1}).$$

The minus signs are conventionally put above the numbers and are said 'bar one' (or 'one bar' if you are American!). You will see that in the cube the six faces are labelled by all the possible combinations of (100). The set of faces is called the $\{100\}$ **form** of the faces (note the convention of curly brackets to represent the *set* of planes consistent with the crystal symmetry). If the cube were pulled out or squashed along **c** the crystal would be tetragonal, and therefore to be consistent with the symmetry, there would be two forms, $\{100\}$ and $\{001\}$, containing the faces

$$\{100\}: (100) \quad (010) \quad (\bar{1}00) \quad (0\bar{1}0)$$

$$\{001\}: (001) \quad (00\bar{1}).$$

Consider now the eight faces of an octahedron (figure 1.6(b)). Each face cuts off equal intercepts along the three cube axes and so, reducing the intercepts to a common factor, the form is {111}, with the eight faces labelled according to all possible combinations:

$$(111) \quad (\bar{1}11) \quad (1\bar{1}1) \quad (11\bar{1}) \quad (1\bar{1}\bar{1}) \quad (\bar{1}1\bar{1}) \quad (\bar{1}\bar{1}1) \quad (\bar{1}\bar{1}\bar{1}).$$

The four faces of a tetrahedron are a subset of the octahedron faces.

Problem 1.5 *Suppose the octahedron were stretched along the $\pm c$ directions. Index the faces and forms.*

Although the crystal is now tetragonal, it should be clear that the indices of the faces are unchanged. In this particular case, the form is also unchanged.

Finally, take the example of iron pyrites in figure 1.6(c). The index (210) is obtained like this. The (210) face makes an intercept halfway along a and makes a unit intercept along b. The face is also parallel to c. Taking common factors, the intercepts are then

$$\frac{a}{2} \quad \frac{b}{1} \quad \frac{c}{0} = \infty.$$

The remaining faces are found by combinations of these indices to give the form {210}.

Lattices

2

2.1 Introduction

The most fundamental feature of a crystal is that it is made from repeating objects. In fact, it is this repetition that lies at the heart of most of solid state physics. It plays a particularly crucial role in determining thermal and electronic properties. Because of the central importance of the lattice concept, we shall concentrate exclusively on it in this chapter. To do this, let's forget that the crystal actually consists of atoms or molecules (we shall deal with this in the next chapter), but instead focus solely on the *way* in which they are repeated throughout space. We then treat the **lattice** as a mathematical abstraction, which leads to the following definition:

A lattice is an infinite, regular array of points in space.

Notice that no mention is made of atoms or any physical objects, just points in space — no more, no less. This separation of the idea of the 'lattice' from that of the 'crystal structure' has been done deliberately in order to stress a very important matter. It must be understood that, strictly speaking, there is no lattice inside a crystal: if we could look through a powerful microscope at a crystal structure we should not expect to see **lattice points**, only atoms or groups of atoms. Do not confuse lattice points with atoms (this is a common mistake, made even by reputable scientists)! The lattice provides the 'recipe' that determines how the atomic or molecular units are to be repeated throughout space to make the crystal structure.

Figure 2.1 shows a section of a typical lattice. Each dot represents a lattice point (mathematically described by a delta function). We choose an origin at any one lattice point, and draw vectors a, b and c to three other lattice points, such that the vectors are neither mutually collinear nor coplanar. Then all the other lattice points lie at the ends of vectors given by the formula

$$t_n = n_1 a + n_2 b + n_3 c. \qquad (2.1)$$

The set of vectors t_n describes the so called **translational symmetry** of a lattice, once an appropriate set of a-, b- and c-axes has been selected. If we

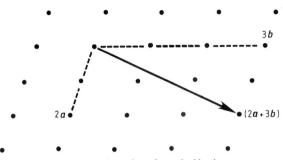

Figure 2.1   A section of a typical lattice.

do it in such a way that the quantities n_i are integers, the translation vectors are said to be **primitive translation vectors**. In the example shown in figure 2.1, we can see one such vector made by going two units along a, three units along b and no units along c. Notice too an important property of a lattice: no matter which lattice point we choose, the environment around the point always looks the same. It is a good idea, when looking at an array of points, to use this criterion to check whether you are dealing with a true lattice.

2.2 Unit Cells

As we have seen above, the lattice is by definition infinite, and clearly this is going to cause us some problems, especially when we come to discuss crystal structures, since it would be tiresome to have to describe the positions of all the lattice points or atoms out to infinity! To get round this difficulty, we can use the notion of translational symmetry to do the work for us. This leads to another definition:

A unit cell is a region of space which when repeated by primitive translation vectors fills all space.

From this it follows that, by considering a **unit cell** on its own, we automatically know all about the rest of the lattice (the concept of 'unit cell' applies equally to crystal structures: there the cell contains a set of atoms to be repeated according to the translational symmetry).

Figure 2.2 shows our original lattice once again, but this time some different types of unit cell have been marked on it. Region A defines one with the smallest possible volume (assuming the third dimension to be treated similarly). Unit cell B, although of a different shape, has the same volume. In fact, there is an infinite choice of such cells of minimum volume.

Problem 2.1 *Draw some more for yourself to convince yourself of this.*

Region C, on the other hand, is twice as large as A or B: it is still a unit cell, as when stacked together with identical copies it reproduces the whole lattice. It should be clear that we can continue to divide up the lattice into units of any number of different shapes and sizes.

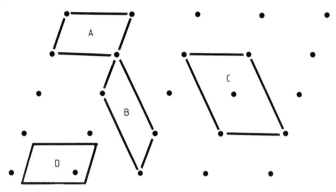

Figure 2.2   Some examples of different choices of unit cell.

Region D has been drawn to emphasise another important idea, namely that the origin of the unit cell is a matter of choice. Thus D is a unit cell of the same shape, size and orientation as A, only its origin has been deliberately shifted off a lattice point. Nevertheless, I hope it is obvious that stacking together identical copies of this cell still produces the whole lattice as before. Furthermore, by drawing the unit cell in this way, we can easily see one lattice point inside it. Unit cells like A, B and D, because they contain just one lattice point, are called **primitive unit cells** (this is better than the term 'simple', which you will often find in textbooks), and enclose the smallest possible repeating unit in a lattice. A primitive cell, or the lattice described in terms of primitive cells, is denoted by the letter P. This idea of shifting the unit cell origin is a neat device for counting the number of lattice points per unit cell (also for atoms per unit cell in a crystal structure), and is much easier than the usual method given in textbooks, where fractions of a lattice point are divided between neighbouring unit cells.

Problem 2.2 *Displace unit cell C slightly away from a lattice point. How many points are there inside?*

I hope you found two. Such a cell, with more than one lattice point, is non-primitive, and in this case must have twice the volume of a primitive cell.

2.3 Lattice Types

From the above discussion, it should be clear that in any lattice we can always define primitive unit cells to describe the whole lattice. However, we can equally well define non-primitive cells as well. This prompts the question: is there any reason why we might *prefer* to choose a non-primitive cell? The answer to this can be seen if we take an example.

Figure 2.3 shows a perspective drawing of a cubic unit cell (remember the four threefold axes of the cubic system) with a lattice point not only at each corner, but also at the centre of each face.

Problem 2.3 *Check by imagining the origin to be displaced that there are four lattice points per cell of this type.*

A smaller, primitive unit cell, in the shape of a rhombohedron, and constructed from this non-primitive cell, is marked in full lines. As this primitive cell contains only one lattice point, it has one quarter of the volume of the cube-shaped cell, and so the advantage of using it when we describe the crystal structure will be that we shall need to specify only one quarter of the number of atoms. Now, suppose we remove this primitive cell and imagine it in isolation. It would not be easy to see that this particular cell is a very special one, with angles between its edges of precisely 60°. This is a consequence of having constructed it from a cube. However, in general, a rhombohedron can have any angle between its edges, the only requirement being that the edges must be of equal length. More importantly, we cannot, in this isolated primitive cell, spot the four threefold axes of symmetry that tell us that the lattice from which it is derived belongs in the cubic system.

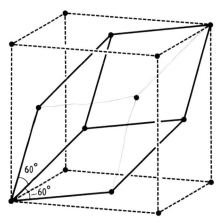

Figure 2.3 The construction of a primitive rhombohedron from a face-centred (F) cubic unit cell.

Therefore, the apparent advantage of using a smaller unit cell is offset by not having obvious symmetry information. It is generally advisable, when describing crystal structures and lattices, to use the smallest possible unit cell that at the same time most directly displays the full symmetry of the lattice.

At this juncture, it is worth considering what sort of non-primitive cells should be used. While there is an infinity of choices, only a limited set are normally chosen by convention. We have seen in figure 2.3 one such useful cell. This one is called a **face-centred** unit cell. This type of cell has lattice points at the positions

$$0, 0, 0 \qquad \tfrac{1}{2}, \tfrac{1}{2}, 0 \qquad \tfrac{1}{2}, 0, \tfrac{1}{2} \qquad 0, \tfrac{1}{2}, \tfrac{1}{2}$$

where the coordinates are specified as fractions of the unit cell edges, with the origin taken on a lattice point. It is conventional to give the symbol F to a lattice described by this type of unit cell.

Figure 2.4(*a*) shows a **body-centred** unit cell, which here, to emphasise that we do not confine ourselves to the cubic system alone, is drawn in the orthorhombic system. There are two lattice points per body-centred cell with fractional coordinates

$$0, 0, 0 \qquad \tfrac{1}{2}, \tfrac{1}{2}, \tfrac{1}{2}.$$

The letter I (from the German *Innenzentrierte*) denotes a body-centred lattice.

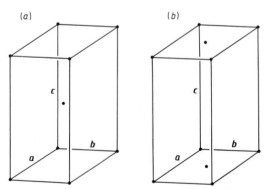

Figure 2.4   (*a*) A body-centred (I) unit cell. (*b*) A C-centred unit cell.

Figure 2.4(*b*) illustrates how only one face per unit cell can be centred. In this example, the faces perpendicular to the *c*-axis are centred, the so called **C-centring.** We could equally choose A- or B-centring. Again, this type of centring creates two lattice points per unit cell, which in C-centring are given by

$$0, 0, 0 \qquad \tfrac{1}{2}, \tfrac{1}{2}, 0.$$

What happens if we try to centre two independent faces of a unit cell? In figure 2.5 an attempt at this is made. Notice that the environment around the points marked A is quite different from that around B (the broken lines indicate this). This cannot form a lattice, since the view from any one point must be the same as from any other. It is generally true that it is not possible to produce a lattice consisting of centring on two independent unit cell faces.

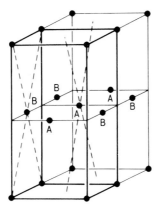

Figure 2.5 A fruitless attempt at centring two independent faces of a unit cell (from Burns and Glazer).

2.4 Bravais Lattices

Suppose that we take one of the crystal systems, say tetragonal, and ask how many different lattice types can be constructed. The bold lines in figure 2.6(a) show a primitive tetragonal unit cell. We see that a larger C-centred

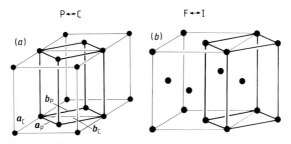

Figure 2.6 (a) A primitive tetragonal unit cell constructed from C-centred tetragonal cells. (b) A body-centred tetragonal cell constructed from a face-centred tetragonal cell (from Burns and Glazer).

cell can be drawn (in faint lines), that also displays the tetragonal symmetry (in this case given by a fourfold axis through the C face). This simply means that in the tetragonal system, C-centred unit cells can be described as primitive cells by redefining the axes, without any loss of essential symmetry information. The two types of cell are not uniquely different. The same relationship is found between I and F cells (figure 2.6(*b*)). However, there is no way to transform a P tetragonal cell to, say, an I tetragonal cell, as they are uniquely different. Thus in the tetragonal system there are two unique types of lattice. It is a matter of taste as to whether we prefer to use a P cell rather than a C cell, or an I cell rather than an F cell, although it is more usual to use the smaller cell. It is a simple matter to express the C-unit cell axes a_C, b_C, c_C in terms of the primitive axes a_P, b_P, c_P by matrices

$$\begin{pmatrix} a_C \\ b_C \\ c_C \end{pmatrix} = \begin{pmatrix} 1 & -1 & 0 \\ 1 & 1 & 0 \\ 0 & 0 & 1 \end{pmatrix} \begin{pmatrix} a_P \\ b_P \\ c_P \end{pmatrix} \tag{2.2}$$

with a similar relationship between the F and I cells. Similar considerations apply to all the other crystal systems. In fact, throughout the seven crystal systems there are fourteen unique lattice types possible. These are known as the **Bravais lattices** or **space lattices**. The easiest way of deriving them is to look in each system for the smallest unit cell, which at the same time shows the full lattice symmetry. Figure 2.7 shows drawings of the 14 space lattices derived in this way.

In the cubic system, there are three Bravais lattices. Obviously, there is always the possibility of a primitive lattice in each crystal system. However, in addition here, we can have face-centring and body-centring. By examining the drawings in figure 2.7 you should be able to see that in each case the four threefold axes of the cubic system are preserved.

Problem 2.4 *Demonstrate with the aid of a diagram that C-centring destroys the four threefold axes of a primitive cube.*

Notice too that in the triclinic system there can only be primitive unit cells, as such cells automatically display the full symmetry, such as it is, of this system.

2.5 Limitations in Symmetry Elements of Lattices

Consider two lattice points A and A' separated by a unit translation t (figure 2.8). Suppose that we apply some symmetry operation, such as a rotation, to the vector t. If the rotation axis is taken to be at A, t will be rotated anticlockwise through an angle α to form a new lattice vector AB.

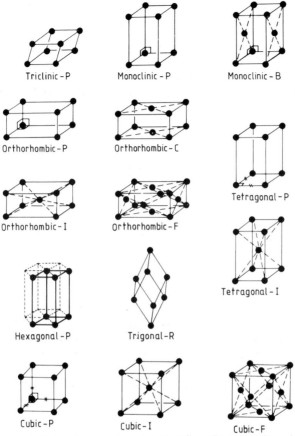

Figure 2.7 The fourteen Bravais lattices (from Burns and Glazer).

Furthermore, as A and A' are lattice points, the same type of rotation operation should apply also at A'. You will recall from chapter 1 that to every operation in a group, there must be an inverse, and so let us just apply the inverse rotation at A'. This rotates A'A clockwise through the angle α to form A'B'. Since B and B' must also be lattice points, the distance BB' must end up as an integral number of basic translations t. We can now write

$$BB' = nt \qquad (2.3)$$

where n is some integer, and from figure 2.8, we find

$$BB' = -2t \cos \alpha + t \qquad (2.4)$$

and combining equations (2.3) and (2.4)

$$\cos \alpha = (1 - n)/2. \qquad (2.5)$$

Now, since n is an integer, let us write $1-n = N$, where N is also an integer. In addition, the angle α must lie somewhere between 0 and 180° in order to ensure that repeated application of the rotation eventually ends up with the translation vector back in the original direction AA'. This means that $\cos\alpha$ must lie between ±1. Therefore

$$|\cos \alpha| \leq 1 \qquad (2.6)$$

and so

$$|N| \leq 2 \qquad (2.7)$$

to give

$$N = -2, -1, 0, 1 \text{ or } 2. \qquad (2.8)$$

These values of N correspond to possible rotations through angles of π, $2\pi/3$, $\pi/2$, $\pi/3$ or 0 radians.

The implication of this is that the only rotational operations that are allowed in a lattice are 2, 3, 4, 6 or 1. This limited set of possible rotations is the reason why there are only seven crystal systems, fourteen Bravais lattices and thirty two crystal classes. It is impossible therefore to construct a lattice with fivefold or sevenfold rotational symmetry. Try it for yourself, by taking several British fifty pence pieces, or, if like me you can't afford them, twenty pence pieces, and see if you can join them together in a plane without leaving any gaps between. Incidentally, the gaps between the coins allow them to be picked up easily from a shop counter, and so these shapes are quite a sensible choice for coins.

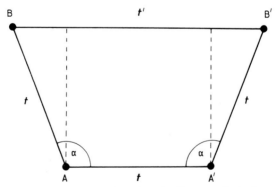

Figure 2.8   The effect of rotating a lattice vector through an angle (from Burns and Glazer).

2.6 Directions and Zones

We saw in chapter 1 that the faces and planes in a crystal could be labelled in terms of the intercepts made on the crystallographic axes. Here, we shall deal with the specification of crystal directions. This is simply denoted by the symbol [*uvw*], referring to the vector *u**a** + v**b** + w**c***. Thus a crystallographic direction is given by a vector specified in terms of the components *u**a**, v**b*** and *w**c*** along the crystal axes. Notice that the convention is to use square brackets to indicate directions. The set of all directions conforming to the symmetry, rather like the form of planes, is conventionally labelled ⟨*uvw*⟩. Do not confuse direction symbols with the Miller indices of planes. Figure 2.9 shows the projection of a unit cell with some directions and planes marked on it.

Problem 2.5 *Study this diagram closely and make sure that you understand how the planes and directions are defined.*

Note that the [100] direction is synonymous with the *a*-axis direction. You can also see that the [110] direction, say, is not perpendicular to the (110) plane. Only in the cubic system is it true to say that [*uvw*] and (*hkl*) are always perpendicular to each other when $u = h$, $v = k$ and $w = l$. Since the direction [*uvw*] is a line in space, it also corresponds to the intersection of a set of planes (figure 2.10), and this set of planes is said to form a **zone** labelled [*uvw*]. The [*uvw*] direction is also called the **zone axis**. Thus all planes of the type (*hk*0) are parallel to the [001] direction and therefore intersect to form a line in this direction. They are said to form the [001] zone.

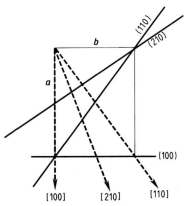

Figure 2.9 A projected unit cell with planes (bold lines) and directions (arrowed). For convenience, the planes are assumed parallel to *c* and the directions are taken in the *ab*-plane.

Problem 2.6   *Which planes belong to the [111] zone?*

These must all be planes that are parallel to [111], and so you should be able to convince yourself that planes such as

$$(\bar{1}01), \quad (10\bar{1}), \quad (\bar{1}10), \quad (\bar{2}11), \quad (2\bar{1}\bar{1}), \quad (1\bar{2}1), \quad (\bar{1}2\bar{1})\dots$$

all intersect on this direction. Actually, an easy way of working this out is as follows. Just write

$$hu + kv + lw = 0 \tag{2.9}$$

to get the relations between h, k and l. Thus for [111], we write

$$h + k + l = 0 \tag{2.10}$$

and so all planes with Miller indices conforming to the condition in equation (2.10) must lie in the [111] zone.

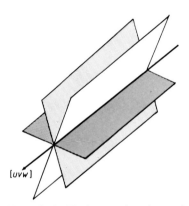

Figure 2.10   The intersection of a set of planes to form a zone axis.

The specification of directions in terms of vectors makes it easy to calculate angles between directions. Thus for the orthorhombic system, the angle Φ between $[uvw]$ and $[u'v'w']$ is given through the use of the scalar product

$$(u\boldsymbol{a} + v\boldsymbol{b} + w\boldsymbol{c})\cdot(u'\boldsymbol{a} + v'\boldsymbol{b} + w'\boldsymbol{c}) = (uu'a^2 + vv'b^2 + ww'c^2) \tag{2.11}$$

which is also equal to

$$(u^2a^2 + v^2b^2 + w^2c^2)^{1/2}(u'^2a^2 + v'^2b^2 + w'^2c^2)^{1/2}\cos\Phi \tag{2.12}$$

and by equating (2.11) to (2.12) it is possible to solve for Φ.

Problem 2.7 *Show that in cubic crystals the angle between any two directions is given by*

$$\cos \Phi = \frac{uu' + vv' + ww'}{[(u^2 + v^2 + w^2)^{1/2} \, (u'^2 + v'^2 + w'^2)^{1/2}].} \qquad (2.13)$$

Thus, for example, the angle between [111] and [100] in a cubic crystal is

$$\cos^{-1} (1/\sqrt{3}) = 54.74°$$

the angle between a cube edge and a body diagonal. Because in cubic crystals any direction is perpendicular to the plane with the same indices, this automatically gives us the angle between plane normals too. The method can easily be generalised to work even in the triclinic system, although the formula is somewhat longer. How nice it is not to have to draw diagrams nor to have to use geometry to work out these angles!

Crystal Structures

3

3.1 Definition of a Crystal Structure

A crystal does not actually consist of lattice points, but rather of repeating objects of some kind. In reality, the **object** consists of a group of atoms, or possibly of a single atom. In principle, however, any object repeated throughout space could be said to form a 'crystal' in the general sense of the word. In many books, the object is called the **basis** of the crystal structure, but you should be warned that you may occasionally see this word used differently, as in the recent edition of the International Tables for Crystallography, where the word 'basis' is used to denote the base vectors a, b and c of the lattice.

How, then, can we use the idea of the lattice and the basis to define a crystal in a mathematical sense? Fortunately, there is a simple but elegant mathematical operation that does this for us. It is known as **convolution**. Suppose that we have two functions of position $f(r)$ and $g(r)$. The convolution of these functions is defined to be

$$h(r) = f(r) * g(r) = \int f(r-r')g(r)dr'. \qquad (3.1)$$

Figure 3.1 shows what this definition actually means with a simple one-dimensional example. I have taken the function $f(r)$ to be some general shape (figure 3.1(a)) and the function $g(r)$ to constitute a set of three sharp line functions of different heights (figure 3.1(b)). The effect of convolution is to slide the first function along the r-direction, forming $f(r-r')$ in the above equation, multiplying with $g(r)$ and integrating as we go. The result of this (figure 3.1(c)) is a set of three copies of the original function $f(r)$, separated by the distances between the three line functions and scaled according to their heights. It is this copying action that makes convolution useful to us in defining the crystal. Let the function $L(r)$ describe the lattice (mathematically, a function that produces a regular array of delta functions), and let $B(r)$ represent the basis. In a real crystal, of course, $B(r)$ would refer to an atomic density function, but since we are being quite general for the moment, we can take $B(r)$ to represent any physical object. Figure 3.2 shows the way the crystal structure can be derived through convolution. For a basis, I have chosen in 3.2(a) a particular object, a shark. In 3.2(b) a typical lattice is given, which we assume to continue in all directions out to infinity. Now, to form the convolution $B(r) * L(r)$, the shark must be slid over the whole of the lattice in (a), multiplying by the delta functions and integrating at the same time. Since the delta functions are by definition zero, except at the points of the lattice, the effect of this is to make an infinite number of identical copies of the shark in sympathy with the periodicity of the lattice (figure 3.2(c)). The result is an infinite crystal $C(r)$ made entirely of sharks!

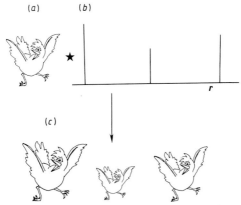

Figure 3.1 An example of convolution: (a), the function $f(r)$; (b), the function $g(r)$; (c), the convolution $f(r) * g(r)$.

We can go even further with this idea. A real crystal, of course, is not infinite in extent, but has a surface. Mathematically, we can describe this by using a 'shape' function $S(r)$, which is simply a function that has the value zero outside a defined region of space and unity within. Figure 3.2(d) shows an arbitrarily chosen shape function. If we use the operation of multiplication thus:

$$C(r) = [B(r) * L(r)] \times S(r)$$

we effectively have a mathematical definition of a finite crystal (figure 3.2(e)).

Put into words, we can make the following definition:

A crystal is a regular and finite array of objects.

This is a very general definition, but of course when the 'objects' in question are groups of atoms, rather than, say, sharks, the definition is consistent with what we normally call a crystal!

In this book, I have been careful to separate the ideas of lattice and crystal, because there is so much confusion about the distinction in many textbooks and among practicing scientists. Do not confuse a lattice point with an atom. Many simple crystal structures have a single atom in the basis and the result is that the drawing of the structure resembles very closely the equivalent drawing of the lattice, since there is then one atom per lattice point. This distinction should become clearer when we discuss in detail actual crystal structures. Do not use the term 'lattice structure': this is an old term commonly used before the discovery of x-ray diffraction made it possible to derive crystal structures. Unfortunately, this term is still very much used, but it is to be discouraged because it leads to confusion between lattices and crystal structures.

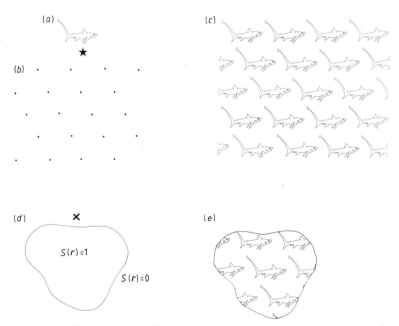

Figure 3.2 The mathematical formation of a crystal structure shown schematically: (*a*), the basis function $B(\mathbf{r})$; (*b*), the lattice function $L(\mathbf{r})$; (*c*), an infinite crystal formed by $B(\mathbf{r}) * L(\mathbf{r})$; (*d*), a shape function $S(\mathbf{r})$; (*e*), a finite crystal formed by multiplying by $S(\mathbf{r})$.

3.2 Examples of Crystal Structures

In this section, we shall use all the ideas that we have met so far to describe several simple, but common crystal structures. We shall see that by the use of symmetry and the lattice concept, a complete structure can be defined economically by a small number of parameters.

3.2.1 Copper

Figure 3.3 shows a unit cell of the crystal structure of metallic copper, both in perspective (*a*) and in projection on the (001) plane (*b*). The figures marked on the projection give the heights of the atoms as fractions of the *c*-axis length. You should get used to using projections like this, as they are much easier to draw than perspective diagrams.

Problem 3.1 *How many copper atoms are there in the unit cell? (Remember the trick of displacing the unit cell origin.)*

You should have found four copper atoms. Now the crystal system of copper is cubic and the lattice type is all-face-centred F. We can thus describe the structure by

system: cubic
lattice: F
basis: Cu at (0, 0, 0).

We can now use this description to build the complete crystal structure. Using the idea of convolution, we start with the basis; here it is a single atom at the position (0, 0, 0,). This single-atom basis is then repeated throughout space in accordance with the F lattice. For a single unit cell, this means that to the basis we must add the **fractional coordinates** (i.e. coordinates denoted as fractions of the unit cell edges)

$$(0, 0, 0) \qquad (\tfrac{1}{2}, \tfrac{1}{2}, 0) \qquad (\tfrac{1}{2}, 0, \tfrac{1}{2}) \qquad (0, \tfrac{1}{2}, \tfrac{1}{2})$$

and, since in the case of copper the basis consists of the single copper atom at (0, 0, 0), the result is a copper atom at each of the four coordinate points of the lattice in the unit cell. Note that the drawing of the copper structure resembles closely that of the cubic F Bravais lattice in figure 2.7. Such crystals, with one atom per lattice point, are sometimes called **Bravais crystals** because of their similarity to the Bravais lattice drawings. In my view, this is an unnecessary term, and again reveals a lack of understanding of the distinction between lattice points and atoms.

Let us find out if this structure is centrosymmetric. To do this, we look for a relationship between the atoms of the type $(x, y, z) \to (-x, -y, -z)$. We can see immediately that the copper atom at, say, $(\tfrac{1}{2}, \tfrac{1}{2}, 0)$ is related by the lattice translation $(-\boldsymbol{a} -\boldsymbol{b})$ to one in a neighbouring unit cell at $(-\tfrac{1}{2}, -\tfrac{1}{2}, 0)$,

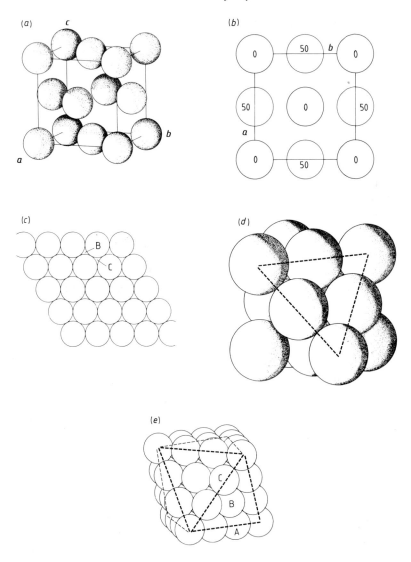

Figure 3.3 (*a*) Perspective view of the crystal structure of metallic copper. (*b*) Projection on (001). The fractional heights of the atoms are marked in hundredths of the *c*-axis. (*c*) A layer of close-packed spheres. (*d*) The structure of copper drawn with the atoms in close contact. One of the corner atoms has been removed to show that the (111) plane, sketched in broken lines, is close packed. (*e*) Cubic close packing of spheres in such a way as to form an octahedron. The drawing has the stacking direction vertical so that the top face is one of the (111) faces of the octahedron.

and this fulfills the necessary condition. Since there is only one atom in the basis, all the other atoms in the unit cell must obey the same type of condition. Thus each copper atom in the unit cell lies on a centre of symmetry. Notice that there can be more than one centre of symmetry in a unit cell, as in this case.

The copper structure is typical of many elements, and because it is built from single atoms at the positions of a **face-centred cubic** array, this structure type is often called simply **fcc**. Another way of considering this structure is through the close packing of spheres. Figure 3.3(c) shows a single layer of close-packed spheres, which we shall denote as an A layer, with two types of interstice between the spheres marked B and C. Now imagine putting another close-packed layer of spheres on top, in such a way that the spheres in the upper layer sit directly on the interstices in the lower layer.

Problem 3.2 *Do the spheres in the upper layer sit directly over both B and C interstices?*

If you sketch the next layer over the lower layer (or construct the layers with ping-pong balls stuck together with glue) you should find that only one type of interstice is covered, say B. We call this upper layer a B layer. The same choice confronts us when we add a third layer. It turns out that if we stack a sequence of such layers in the order ABCABCABCABC . . . (the stacking direction), the resulting structure is fcc, with the close-packed layers forming (111) planes. Figure 3.3(d) illustrates the copper structure seen as a **cubic close-packed** array. Another way of seeing this is to cut the boundaries of the close-packed layers in such a way that they form an octahedron (figure 3.3(e)), each face of which consists of a close-packed layer. Remember that the faces of an octahedron are cubic {111} faces. Table 3.1 gives the fraction (**packing fraction**) of the total unit cell volume occupied by atoms, assuming an ideal packing of spheres; we see that approximately 74 % of the unit cell volume is occupied by spherical atoms, and from the measured cell constant we can use this to estimate the effective atomic radius.

Table 3.1 Packing fractions of different structure types.

fcc	0.74	hcp	0.74
simple cubic	0.52	bcc	0.68
diamond	0.34		

3.2.2 Diamond and Zinc Blende

Diamond, silicon and germanium adopt similar types of crystal structure. This structure is possibly the most important one of recent times because of its relevance to the silicon chip technology of today. In order to understand the physical properties of these materials, it is essential that the solid state physicist should have a full understanding of the crystal structure. The diamond structure is shown in perspective and in projection in figure 3.4.

Problem 3.3   *How many carbon atoms are there in the unit cell of diamond?*

If you displace the unit cell origin in figure 3.4(*b*) you should be able to count eight. The description of the structure is:

system: cubic
lattice: F
basis: C at $(0, 0, 0)$
             C at $(\frac{1}{4}, \frac{1}{4}, \frac{1}{4})$.

In this case, the basis consists of two identical atoms. Applying the convolution principle to each atom in the basis we now add each coordinate of the F lattice:

Atoms	Lattice	Result
$(0, 0, 0)$	$+ (0, 0, 0) =$	$(0, 0, 0)$
$(0, 0, 0)$	$+ (\frac{1}{2}, \frac{1}{2}, 0) =$	$(\frac{1}{2}, \frac{1}{2}, 0)$
$(0, 0, 0)$	$+ (\frac{1}{2}, 0, \frac{1}{2}) =$	$(\frac{1}{2}, 0, \frac{1}{2})$
$(0, 0, 0)$	$+ (0, \frac{1}{2}, \frac{1}{2}) =$	$(0, \frac{1}{2}, \frac{1}{2})$
$(\frac{1}{4}, \frac{1}{4}, \frac{1}{4})$	$+ (0, 0, 0) =$	$(\frac{1}{4}, \frac{1}{4}, \frac{1}{4})$
$(\frac{1}{4}, \frac{1}{4}, \frac{1}{4})$	$+ (\frac{1}{2}, \frac{1}{2}, 0) =$	$(\frac{3}{4}, \frac{3}{4}, 0)$
$(\frac{1}{4}, \frac{1}{4}, \frac{1}{4})$	$+ (\frac{1}{2}, 0, \frac{1}{2}) =$	$(\frac{3}{4}, 0, \frac{3}{4})$
$(\frac{1}{4}, \frac{1}{4}, \frac{1}{4})$	$+ (0, \frac{1}{2}, \frac{1}{2}) =$	$(\frac{1}{4}, \frac{3}{4}, \frac{3}{4})$

This results in eight carbon atoms arranged as in the diagram. It can be seen that each carbon atom is at the centre of a tetrahedron of other carbon atoms to form a compact structure. For example, the carbon at $(\frac{1}{4}, \frac{1}{4}, \frac{1}{4})$ is at the

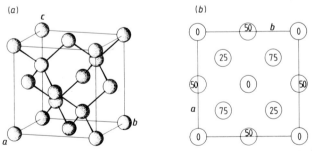

Figure 3.4 (*a*) Perspective view of the diamond structure. The shortest C–C bonds have been drawn to show the tetrahedra. (*b*) Projection on (001).

centre of a tetrahedron described by the two atoms at $(0, 0, 0)$ and $(\frac{1}{2}, \frac{1}{2}, 0)$ below, and the two atoms at $(\frac{1}{2}, 0, \frac{1}{2})$ and $(0, \frac{1}{2}, \frac{1}{2})$ above.

Problem 3.4 *Draw in lines on figure 3.4(b) to join these four atoms and verify for yourself that they form a tetrahedron.*

If we assume close-packing of hard spheres, we determine that this structure is only approximately 34 % occupied by the atomic volume (table 3.1). You may care to consider how it is that one of the hardest substances known should have so much empty space in it. The clue comes from the fact that carbon is very small compared with the other elements.

Notice too, that this time, in contrast to copper, the structure no longer resembles any of the drawings of the Bravais lattices, since for each lattice point in the cubic F Bravais lattices there are now two carbon atoms. Let me warn you too that, in many books and research papers, you will see the term 'diamond lattice' used, when what is really meant is 'diamond structure'. The diamond lattice, strictly speaking, is identical to the copper lattice (apart from its actual dimensions) i.e. it is a cubic F lattice. The diamond structure, on the other hand, is physically different. I hope that after reading this book you will not make this type of mistake!

Let us, as we did for the structure of copper, look for a centre of symmetry. This time, it is not so obvious. First, see if the carbon atom sits on a centre. If it did, we should then expect to find the carbon at $(\frac{1}{4}, \frac{1}{4}, \frac{1}{4})$ related to one outside the unit cell at $(-\frac{1}{4}, -\frac{1}{4}, -\frac{1}{4})$. This second position is equivalent by lattice translations to the position at $(\frac{3}{4}, \frac{3}{4}, \frac{3}{4})$ within the unit cell; however, from the above list of atom coordinates we see that this position is not occupied by any atom, and so there cannot be a centre of symmetry located at a carbon atom. Let us therefore try another place. Suppose that we test for a centre of symmetry at the point halfway along a C–C bond. For instance, there is one such bond between the carbon at $(0, 0, 0)$ and another at $(\frac{1}{4}, \frac{1}{4}, \frac{1}{4})$, so that our 'test point' will be at $(\frac{1}{8}, \frac{1}{8}, \frac{1}{8})$.

Problem 3.5 *Subtract $(\frac{1}{8}, \frac{1}{8}, \frac{1}{8})$ from the coordinates of all the carbon atoms in the unit cell to produce a new list of coordinates specified with respect to our 'test point'. Does this new list suggest a centre of symmetry at the 'test point'?*

You should have found that the eight coordinates divide into two sets of four, with each set related by $(x, y, z) \rightarrow (-x, -y, -z)$, provided that due account is taken of lattice translations, and so the point halfway along a C–C bond does indeed turn out to be a centre of symmetry.

Problem 3.6 *To form the zinc blende (ZnS) structure, replace one carbon in the basis by a zinc atom and the other by a sulphur. Draw a projection on (001). Is this structure centrosymmetric?*

This time, no matter where you search, you should not be able to find a centre of symmetry. Thus ZnS has a non-centrosymmetric structure, and in fact this is seen in its physical properties. For instance, it develops an electric charge when heated, the so called pyroelectric effect. Since the production of a charge means that one face of the crystal must be positively charged and the opposite face negatively charged, this property is also a non-centrosymmetric property, and therefore can only occur in non-centrosymmetric crystal structures. ZnS develops electric charges also when squeezed, the piezoelectric effect, for similar reasons. Diamond, on the other hand, does not have such properties. Several other technologically important substances crystallise in the zinc blende structure, notably GaAs and InSb, which we can expect to hear more about in the near future.

Finally, let me suggest another way of constructing the diamond and zinc blende structures which can be useful. Substitute each atom in a cubic close-packed array by the pair of atoms in the basis given above, so that the vector separating the atoms in the pair is along the stacking direction. Try it yourself with ping-pong balls and you will learn a great deal about the diamond structure.

3.2.3 *Rock Salt*

The rock salt structure is common to many alkali halides, e.g. NaCl, and is shown in figure 3.5. The description of the structure is:

system: cubic
lattice: F
basis: Na at $(0, 0, 0)$
            Cl at $(\frac{1}{2}, 0, 0)$.

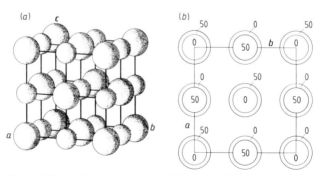

Figure 3.5 (*a*) Perspective view of the rock salt structure. Large circles are sodium atoms; small circles are chlorine atoms. (*b*) Projection on (001).

Once again, applying all the lattice coordinates in turn to each atom in the basis we get four sodiums and four chlorines in the unit cell

Na at $(0, 0, 0)$, $(\frac{1}{2}, \frac{1}{2}, 0)$, $(\frac{1}{2}, 0, \frac{1}{2})$, $(0, \frac{1}{2}, \frac{1}{2})$
Cl at $(\frac{1}{2}, 0, 0)$, $(0, \frac{1}{2}, 0)$, $(0, 0, \frac{1}{2})$, $(\frac{1}{2}, \frac{1}{2}, \frac{1}{2})$.

Problem 3.7   *Choose a different set of atoms to represent the basis, e.g. Na at $(\frac{1}{2}, \frac{1}{2}, 0)$ and Cl at $(\frac{1}{2}, \frac{1}{2}, \frac{1}{2})$, and show that convolution with the lattice gives the same set of atomic positions in the unit cell.*

Note that once again, the sodium chloride lattice is the same lattice, apart from physical size, as the diamond lattice or the copper lattice, but the structure is totally different.

What about a centre of symmetry? This one is easy, as we can see that there is a Cl at $(\frac{1}{2}, \frac{1}{2}, \frac{1}{2})$ related to a Cl at $(-\frac{1}{2}, -\frac{1}{2}, -\frac{1}{2})$ by the lattice translation $(-\boldsymbol{a} -\boldsymbol{b} -\boldsymbol{c})$. Similarly, the Na at $(\frac{1}{2}, \frac{1}{2}, 0)$ is related to a Na at $(-\frac{1}{2}, -\frac{1}{2}, 0)$ by the same lattice translation. Thus, there are centres of symmetry at both Na and Cl positions, and we therefore would not expect alkali halides like these to exhibit polar physical properties.

3.2.4 *Fluorite*

The fluorite (CaF_2) structure is closely related to the ZnS structure. Its description is:

system: cubic
lattice: F
basis: Ca at $(0, 0, 0)$
           F at $(\frac{1}{4}, \frac{1}{4}, \frac{1}{4})$
           F at $(-\frac{1}{4}, -\frac{1}{4}, -\frac{1}{4})$

and the structure is shown in figure 3.6. This time, we observe that each Ca atom is coordinated by eight F atoms and each F atom by four Ca atoms.

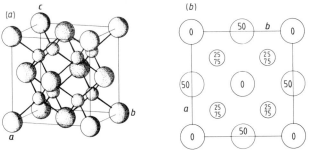

Figure 3.6 (*a*) Perspective view of the fluorite structure. Large circles are calcium atoms; small circles are fluorine atoms. (*b*) Projection on (001).

Problem 3.8 *How many atoms are there in the unit cell?*
How many CaF$_2$ formula units are there per unit cell?
How many atoms are there per lattice point?
Is the structure centrosymmetric?

The first three answers are, in turn, twelve (four lattice points in F centring multiplied by three atoms in the basis); four (the same as the number of lattice points per unit cell, because here the number of atoms in the formula unit is equivalent to the number of atoms in the basis); and three (the number of atoms in the basis). Although the calcium and the first fluorine in the basis build the equivalent of the ZnS structure, the addition of a second fluorine turns it back to being a centrosymmetric structure. Compare figure 3.6(a) with the drawing of the diamond structure in figure 3.4(a) and make sure that you understand the difference between them.

3.2.5 *Iron*

Figure 3.7 shows the iron structure. This is defined by:

> system: cubic
> lattice: I
> basis: Fe at $(0, 0, 0)$.

By this time, you should be getting used to the idea of adding all the lattice vectors to the atoms in the basis. Here, we are back to a one-atom basis, but now with a body-centred lattice. The structure then has two iron atoms in the unit cell at $(0, 0, 0)$ and $(\frac{1}{2}, \frac{1}{2}, \frac{1}{2})$. As in the case of copper, it looks just like the drawing of the cubic I Bravais lattice, and it is centrosymmetric. Many other elements, such as molybdenum, have the same crystal structure, known

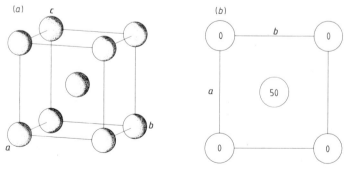

Figure 3.7 (a) Perspective view of the iron structure. (b) Projection on (001).

often as **bcc (body-centred cubic)**. The packing fraction is twice that of the diamond structure.

3.2.6 *Caesium chloride*

Figure 3.8 shows the caesium chloride structure. At first sight it looks similar to that of iron and so we might expect to discover that it is body-centred cubic. However, this would be wrong, since the atom at the body centre of the cubic unit cell is not the same as any of the atoms at the corners. The lattice is therefore primitive, with a two-atom basis:

system: cubic
lattice: P
basis: Cs at $(0, 0, 0)$
          Cl at $(\frac{1}{2}, \frac{1}{2}, \frac{1}{2})$.

Make sure that you understand this fundamental difference between Fe and CsCl, because you will sometimes see statements in textbooks which unfortunately seem to suggest that CsCl is body centred.

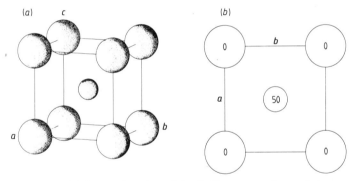

Figure 3.8 (*a*) Perspective view of the CsCl structure. Large circles are caesium atoms; small circles are chlorine atoms. (*b*) Projection on (001).

An interesting structure of this type is that of β-brass, which at low temperatures looks like that of CsCl, with a Cu substituting for the Cs and a Zn for the Cl. It is therefore primitive cubic too. However, on heating β-brass, the Cu and Zn atoms start to hop onto each other's sites to form a disordered structure. Eventually, at a high enough temperature, there are on average as many Cu atoms as Zn atoms on any one site. In this case, although for any particular unit cell we can distinguish the two types of atomic site, when we consider an average over the whole crystal, we cannot distinguish between atoms at the corners and body centres of the unit cells. The lattice then becomes 'effectively' body centred. For instance, an x-ray experiment made at high temperature would give a diffraction pattern typical of a cubic I lattice.

3.2.7 Hexagonal Close Packing

Let us return to the stacking of layers of close-packed spheres. Suppose now that we follow the sequence ABABABAB . . ., thus leaving the C interstices uncovered. The structure thus formed (figure 3.9) is hexagonal, and the structure type is known, therefore, as **hexagonal close packed** or **hcp**. Several elements crystallise in the hcp structure, e.g. beryllium and magnesium, and the structure can be described by:

$$
\begin{aligned}
&\text{system:} \quad &&\text{hexagonal} \\
&\text{lattice:} \quad &&\text{P} \\
&\text{basis:} \quad &&\text{Mg at } (0, 0, 0) \\
& &&\text{Mg at } (\tfrac{1}{3}, \tfrac{2}{3}, \tfrac{1}{2}).
\end{aligned}
$$

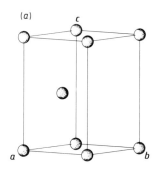

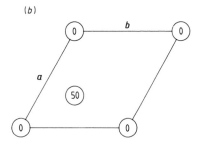

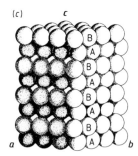

Figure 3.9 (*a*) Perspective view of hexagonal close packing (atoms drawn small to show their positions clearly). (*b*) Projection on (001). (*c*) A large array of hexagonally close-packed spheres.

The packing fraction is identical to that of the fcc structure, and it is for this reason that many metals can transform at certain temperatures between fcc and hcp structures.

Problem 3.9 *Assuming an ideal hcp structure, show that the ratio of the c- to the a-axis is* $\sqrt{(8/3)}$.

The observed *c/a* ratios often show significant departures from this ideal figure, although magnesium comes close to being ideal.

Diffraction by Crystals

4

4.1 Introduction

It is well known that when x-rays are incident on a crystal, a characteristic diffraction pattern of spots is produced on a photographic film placed around the crystal, see e.g. figure 4.1. A full analysis of the diffraction pattern produced with x-rays is highly mathematical and lies outside the scope of this book. However, by using some of the simple ideas of convolution that we met in chapter 3, we can give a good feel for the way in which the pattern is produced. We shall need two new ideas first of all.

When light is diffracted by any object (I use light in its general sense as electromagnetic radiation, and this includes x-rays), the incident wave is broken up into many waves scattered into different directions†. Each scattered wave has its own phase relationship with the incident wave and its own amplitude. I shall assert without proof that the amplitude and phase of the scattered beam can be calculated by taking the so called **Fourier transform**‡ of the scattering object. This is given by the formula:

$$G(s) = \int g(r) \exp(2\pi i r \cdot s) dr. \qquad (4.1)$$

The function $g(r)$ represents the density of the scattering object at a position given by the vector r in the object (for x-rays the density refers to the electron density, as it is the electrons that mainly scatter the x-rays). The function $G(s)$ is the scattered amplitude at a position given by the vector s in the diffraction pattern. Thus this formula takes us from the scattering object, in the so called 'real' space given by r to the diffraction pattern in the 'diffraction' space given by s. The formula can also be reversed (with a negative sign in the exponential term) so that the Fourier transform (FT) of

---

†Refer to C A Taylor 1987 *Diffraction* (Bristol: Adam Hilger).
‡Refer to D C Champeney 1985 *Fourier Transforms in Physics* (Bristol: Adam Hilger).

the diffraction space is real space:

real space (*r*) $\xrightarrow{\text{FT}}$ diffraction space (*s*) $\xrightarrow{\text{FT}}$ real space (*r*).

The calculated amplitude at *s* can then be turned into a scattered intensity by

$$I(s) = G(s) \cdot G(s)^*. \tag{4.2}$$

We shall not get into detailed calculations of diffraction patterns using the Fourier transform, but, for the purposes of our discussion here, we shall note two important results of Fourier transformation: length scales are inverted and symmetry is conserved.

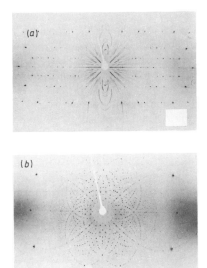

Figure 4.1 (*a*) An x-ray diffraction photograph taken of a crystal of TeO_2 oscillating about the vertical axis (film placed cylindrically around the specimen and monochromatic x-rays used). (*b*) A Laue photograph of a stationary TeO_2 crystal with a polychromatic x-ray beam along the *c*-axis. The crystal is tetragonal and a fourfold axis can be seen on the photograph (allowing for the distortion caused by using a film placed cylindrically around the specimen).

The other important relationship needed for our discussion is what is termed the **Convolution Theorem**, which again I shall state without proof. This simply says that the Fourier transform of the convolution of two functions is equal to the product of their individual Fourier transforms, and vice versa. Thus:

$$\text{FT}[f * g] = \text{FT}[f] \times \text{FT}[g] \qquad (4.3a)$$

$$\text{FT}[f \times g] = \text{FT}[f] * \text{FT}[g]. \qquad (4.3b)$$

We shall now see that these concepts allow us to explain diffraction patterns in a simple manner.

Let us return to our mathematical definition of an infinite crystal:

$$C(r) = B(r) * L(r). \qquad (4.4)$$

Taking the Fourier transform and applying the convolution theorem, we get

$$\text{FT}[B(r) * L(r)] = \text{FT}[B(r)] \times \text{FT}[L(r)]. \qquad (4.5)$$

We can see now that by working out the Fourier transforms of the basis $B(r)$ and the lattice function $L(r)$ separately, the amplitude of scattering by the crystal can be derived. Let us do this in stages.

What is the Fourier transform of a lattice? We could define $L(r)$ mathematically and then perform the integration in equation (4.1): however, we shall do it in a hand-waving way which I think gives a good feel for the calculation. To simplify the argument, consider a single row of lattice points (figure 4.2(a)). To form its Fourier transform, we note first that the lattice has periodicity a, and so in this direction the Fourier transform must also be periodic; but since length scales are inverted, the repeat will now be proportional to $1/a$. In real space, the lattice consists of infinitesimally sharp points, and so in all directions perpendicular to the row of points the Fourier-transformed function must spread out to infinity i.e. it forms a set of repeating infinite planes spaced apart at a distance proportional to $1/a$ (figure 4.2(b)).

Problem 4.1   *How is a three-dimensional lattice related to three single rows of lattice points?*

Figure 4.2(c) shows that if two single rows of lattice points $L(r_1)$ and $L(r_2)$ are convolved together a planar array of points results. Thus to form a three-dimensional lattice we work out

$$L(r) = L(r_1) * L(r_2) * L(r_3) \qquad (4.6)$$

and hence:

$$\text{FT}[L(r)] = \text{FT}[L(r_1)] \times \text{FT}[L(r_2)] \times \text{FT}[L(r_3)]. \qquad (4.7)$$

Now we see the power of the convolution theorem. The Fourier transform of a three-dimensional lattice is simply given by the products of the Fourier transforms of three one-dimensional lattices. Figure 4.2(d) shows this in two dimensions, where we see that multiplication of two sets of repeating infinite

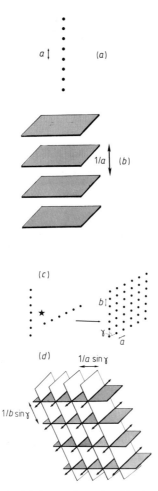

Figure 4.2 (a) A one-dimensional lattice and (b) its Fourier transform. The planes shown are assumed to be infinite in extent. (c) The convolution of two single rows of lattice points to produce a planar lattice, and (d) its Fourier transform (shown slightly inclined) derived as the product of the Fourier transforms of the original single lattice rows. In two dimensions, the multiplication of two sets of planes produces zero everywhere, except on the lines where they intersect (marked by arrows), separated, in this example, by the distances $1/(a \sin\gamma)$ and $1/(b \sin \gamma)$. In three dimensions, there would be three sets of intersecting planes to produce points of intersection i.e. a lattice.

planes gives rise to an infinite regular array of lines of intersection (or points in the case of three dimensions, where three sets of intersecting planes occur). Thus the Fourier transform of a real three-dimensional lattice is also a three-dimensional lattice, but with dimensions reciprocal to those in the real lattice! This new lattice is given a name, originally due to P P Ewald: the **reciprocal lattice**. If we denote the reciprocal axes in this lattice by a^*, b^* and c^*, these are related to the real axes by:

$$a^* = \frac{k(b \cdot c)}{(a \cdot b \times c)} \tag{4.8a}$$

$$b^* = \frac{k(c \cdot a)}{(a \cdot b \times c)} \tag{4.8b}$$

$$c^* = \frac{k(a \cdot b)}{(a \cdot b \times c)} \tag{4.8c}$$

which in turn lead to the conditions:

$$a \cdot a^* = b \cdot b^* = c \cdot c^* = k \tag{4.8d}$$

$$a \cdot b^* = b \cdot c^* = c \cdot a^* = 0. \tag{4.8e}$$

The constant k in equations (4.8) is taken by crystallographers to be equal to one (this is consistent with our definition of the Fourier transform in equation (4.1)) and by physicists to be equal to 2π.

Problem 4.2 *Show that it is also true that* $b \cdot a^* = c \cdot b^* = a \cdot c^* = 0$.

Let us turn our attention to the Fourier transform of the basis. The basis, of course, is some general function representing an object. Suppose that its Fourier transform is as shown in figure 4.3(a). Now, according to equation (4.5), the product of this quantity with the reciprocal lattice will give us the Fourier transform of the infinite crystal. Since the reciprocal lattice consists of empty space populated regularly by points (delta functions), the product of the two will give zeroes between the reciprocal lattice points and infinitesimally narrow peaks of varying height at the nodes of the lattice. That is, the reciprocal lattice samples the underlying transform of the motif (figure 4.3(b)). This can then be converted to an intensity distribution by multiplying the Fourier transform by its complex conjugate. This is essentially the diffraction pattern expected when x-rays are scattered by an infinite crystal.

At this point, I should say a few words about what is meant by 'infinite' within the context of x-ray diffraction. The x-ray wavelengths lie in the region of 0.1–0.2 nm, comparable with the primitive unit cell dimensions in the crystal.

Problem 4.3 *Estimate how many primitive unit cells there are in a crystal of about* $0.2 \times 0.2 \times 0.2$ mm^3 *(a typical size used in modern x-ray diffraction experiments).*

Naturally, you obtained a huge number, and this is what I mean by infinite here. However, suppose that the crystal is only about one micron across, so that the number of primitive unit cells goes down by a factor of 8×10^6: we must then begin to take account of the crystal boundaries. We saw in chapter 3 that a finite crystal was defined by multiplying by a shape function $S(r)$.

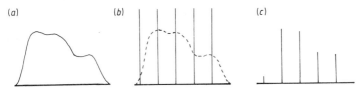

Figure 4.3 (*a*) The Fourier transform of a hypothetical basis and of a lattice. The drawings are shown in one dimension for simplicity. (*b*) and (*c*) The effect of multiplying these transforms together.

Applying the convolution theorem we find that the Fourier transform of a finite crystal (figure 4.4) is given by:

$$\{\text{FT}[B(r)] \times \text{FT}[L(r)]\} * \text{FT}[S(r)].$$

As we make the crystal smaller, the Fourier transform of the shape function grows larger, and since it is convolved with the Fourier transform of the crystal, the effect is to increase the breadth of each of the diffraction peaks. In effect, the shape function is actually a resolution function; the smaller it is, the fewer the number of atoms in the crystal, thus contributing less information to the diffraction pattern and hence making the diffraction pattern less sharp. This peak broadening is in fact used to estimate average particle sizes in materials containing large numbers of small crystallites, for example in alloys.

4.2 The Structure Factor

In this section, we shall learn how to calculate the diffraction intensity from a knowledge of the crystal structure. To do this, we first need a formula for the amplitude of scattering, which can be obtained by Fourier transformation of the electron density in the crystal. After some manipulation, this leads to an important new formula:

$$F(hkl) = \sum_j f_j \exp[2\pi i(hx_j + ky_j + lz_j)]. \qquad (4.9)$$

$F(hkl)$ is known as the **structure factor** and represents the amplitude of scattering by the crystal. In other words, it is the Fourier transform of the crystal. It differs from the function $G(s)$ in that it takes values only at reciprocal lattice nodes given by $s = ha^* + kb^* + lc^*$ and is zero in between. The scattering at particular points in the reciprocal lattice means that on a film placed around the specimen, one obtains sharp spots of scattering, the **Bragg reflections** (so called because Bragg used reflection as an analogue of diffraction in his explanation of x-ray scattering), each one coming from the set of planes of index (hkl). The precise derivation of the structure factor formula can be found in the standard texts on x-ray diffraction. The summation is taken over the j atoms in the unit cell, whose fractional coordinates are x_j, y_j and z_j.

Problem 4.4 *The vector to a reciprocal lattice point is given by $s = ha^* + kb^* + lc^*$ and the vector to the jth atom in the unit cell is given by $r_j = x_j a + y_j b + z_j c$. Using the relationships in equation (4.8), show that $r_j \cdot s = hx_j + ky_j + lz_j$.*

The quantity f_j is known variously as the **form factor**, the **atomic scattering factor**, or particularly in neutron diffraction, the **scattering cross section** or **scattering length** (there are actually small differences in definition between some of these terms, but we shall not concern ourselves with these here). It represents the amplitude of scattering of the individual atoms in the unit cell. Thus, in the case of x-rays, f_j is the amplitude scattered by a ball of electrons. Since we can take the electron distribution around an isolated atom to be spherical, the form factor is then simply the Fourier transform of this ball of

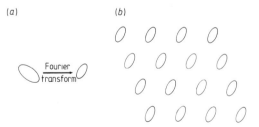

Figure 4.4 The effect of the shape function, shown schematically here in two dimensions. (*a*) The shape function is first Fourier transformed, and (*b*) this is then convoluted with the transform of the infinite crystal to produce broad peaks.

electrons. The Fourier transform of a sphere must itself have spherical symmetry and fall off towards high scattering angles. In the forward direction, each electron scatters the x-rays in phase, so that here the form factor is given by the atomic number of the atom. At other angles, the electrons progressively scatter out of phase to reduce the amplitude of scattering.

Figure 4.5 shows some typical form factor curves plotted as a function of $(\sin \Theta)/\lambda$ where 2Θ is the angle of scattering. Note that atoms with more electrons (higher atomic number) scatter more strongly than those with low atomic numbers. It should be clear from this discussion that, if the electronic wavefunctions ψ for the free atoms are known, the form factor can be calculated by Fourier transformation of the electron density given by $\psi^*\psi$. Generally, the atomic wavefunctions are well known and so the tabulated values of the x-ray form factors are quite accurate. The situation is less well understood in neutron diffraction. Here, it is the nucleus that scatters the neutron. Since this is a tiny object compared with the ball of electrons around an atom, the Fourier transform is large out to high angles, so that to all intents and purposes it can be taken to be independent of angle. Its actual

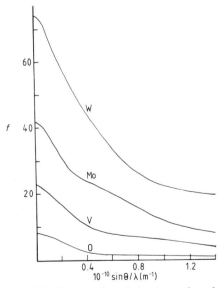

Figure 4.5 The x-ray form factor curves for a few elements.

value depends on the nature of the nucleus and isotope content, and generally it cannot be calculated precisely. As a result, neutron scattering lengths are less well known than the corresponding x-ray form factors.

If the exponential term in equation (4.9) is set equal to one, the structure factor is the amplitude scattered by atoms in a random arrangement; i.e. apart from a scale factor it is the amplitude scattered by a gas of atoms. The intensity is given by multiplying by the complex conjugate, so that the x-ray diffraction pattern from a gas is simply a diffuse intensity, peaked in the forward direction and falling off towards higher angles. The exponential factor is what turns the pattern into an array of spots typical of the diffraction from a crystal, i.e. it incorporates the idea of repetition. As we shall see below, it is also responsible for the absence (**lattice absences**) of certain reflections whose values of h, k and l do not satisfy certain criteria.

Let us now see if we can use the structure factor formula in actual examples. We shall start with the structure of iron. From chapter 3, we see that the structure is body centred with a single Fe atom taken to lie at the origin. Now the structure factor is the Fourier transform of the crystal and so it can be cast in the form:

$$F(hkl) = \text{FT[motif]} \times \text{FT[lattice]}$$

using our old friend, the convolution theorem. In this example, the Fourier transform of the basis is simple. The basis consists of a single atom, and so its Fourier transform is just the form factor f_{Fe}. The lattice is body centred and so its Fourier transform is given by:

$$\exp[2\pi i(0 + 0 + 0)] + \exp[2\pi i(\tfrac{1}{2}h + \tfrac{1}{2}k + \tfrac{1}{2}l)] = 1 + \cos \pi(h + k + l). \quad (4.10)$$

Thus the structure factor is given by

$$F(hkl) = f_{Fe}\{1 + \cos \pi(h + k + l)\}. \quad (4.11)$$

This is equal to zero whenever the sum $h + k + l$ is an odd number, and so such reflections will be absent from the diffraction pattern of iron. A very important point to note is that, because the expression in curly brackets is due to the lattice alone and because it multiplies the Fourier transform of the basis, this condition for absent reflections applies to all crystals with body-centred lattices, irrespective of the crystal system and numbers of atoms.

Now take the diamond structure. The basis consists of a carbon atom at $(0, 0, 0)$ and another at $(\tfrac{1}{4}, \tfrac{1}{4}, \tfrac{1}{4})$ and the lattice is F-centred. The Fourier transform of the basis is:

$$\exp[2\pi i(0 + 0 + 0)] + \exp[2\pi i(\tfrac{1}{4}h + \tfrac{1}{4}k + \tfrac{1}{4}l)]$$
$$= 1 + \cos[\pi(h + k + l)/2] + i\sin[\pi(h + k + l)/2] \quad (4.12)$$

and the Fourier transform of the lattice is:

$$\exp[2\pi i(0 + 0 + 0)] + \exp[2\pi i(\tfrac{1}{2}h + \tfrac{1}{2}k)]$$
$$+ \exp[2\pi i(\tfrac{1}{2}h + \tfrac{1}{2}l)] + \exp[2\pi i(\tfrac{1}{2}k + \tfrac{1}{2}l)]$$
$$= 1 + \cos \pi(h + k) + \cos \pi(h + l) + \cos \pi(k + l). \quad (4.13)$$

Equation (4.13), which applies to all face-centred lattices, gives the condition that reflections with indices of mixed parity must be absent, i.e. h, k and l must be all even or all odd for the reflection to be observed. However, note that in diamond the Fourier transform of the basis (equation (4.12)), which multiplies the lattice transform, also gives rise to certain conditions for absence. In particular, it is zero whenever $h + k + l = 4n + 2$. For example, it predicts that the 222 reflection will be absent, and this is a characteristic feature of the diffraction pattern of diamond. We see, therefore, that there can be absences in the diffraction pattern in addition to those caused by the lattice type alone.

Problem 4.5 *Note that in equation (4.12), an imaginary term appears. This occurs because we have described the structure from an origin that is not on a centre of symmetry. Repeat the structure factor calculation with the coordinates specified with respect to the centre of symmetry and show that only cosine terms appear.*

4.3 The Effect of Temperature

What is the effect of thermal vibrations of the atoms on the diffraction pattern? I guess that your instinctive reply is that it smears out the peaks. However, let us look at the problem more closely.

Suppose, for simplicity, that all the atoms are vibrating independently and isotropically about their mean positions. We do not have to worry about the frequencies of the vibrations as these are typically of the order of 10^{13} s^{-1}, whereas x-ray frequencies are of the order of 10^{19} s^{-1}, and so the energies of the x-rays incident on the crystal will be hardly altered by the energy of the oscillations. Thus x-rays can be taken to undergo elastic scattering by the atoms in the crystal, and so they can only give information about the time-averaged positions of the atoms. Thus, if a powerful x-ray microscope, capable of viewing the crystal structure directly, were available, we should expect each atom to appear fuzzy.

How can the oscillation be represented in real space? We have seen how to define a finite crystal using convolution and multiplication operations. Now we have to alter this by including the effect of the fuzziness of the atoms. This is achieved simply by inventing a 'smearing' function $D(\mathbf{r})$, which is really just another example of a shape function. However, in this case, the smearing function must affect all the unit cells in the structure equally. This, of course, is perfect for the convolution operation, and so we can write a new expression for a crystal with vibrating atoms as:

$$C_{vib}(r) = D(r) * C_{static}(r) \qquad (4.14)$$

and then the diffraction pattern is given by the Fourier transform:

$$G(s) = FT[D(r)] \times FT[C_{static}(r)]. \qquad (4.15)$$

If the vibrational amplitudes are large enough (typical amplitudes in crystals are of the order of 0.01 nm), the Fourier transform of $D(r)$ will be given by a smooth curve that falls off smoothly with scattering angle, rather like an x-ray form factor (indeed if you think about it, the thermal vibration factor is rather like a form factor, as it spreads each atom out, and hence its electron density, to make it effectively larger). Since this now multiplies the Fourier transform of the static crystal, it merely reduces the heights of the diffraction peaks monotonically with increasing scattering angle. No effect on the breadths of the peaks is found.

From this kind of analysis it is possible to modify the structure factor formula to include the effect of thermal vibrations:

$$F(hkl) = \sum_j f_j \exp[-B(\sin^2 \Theta)/\lambda^2]\exp[2\pi i(hx_j + ky_j + lz_j). \qquad (4.16)$$

B is called the overall **temperature factor**, and is related to the amplitude of oscillation u by

$$B = 8\pi \overline{u^2}. \qquad (4.17)$$

Notice that it appears in a negative exponential term, to give the necessary fall-off in amplitude with angle, which multiplies the original structure factor formula, in line with our prediction. Crystallographers these days compute the temperature factors routinely along with the atomic positional parameters as part of their analysis of the diffractometer data. Usually, temperature factors of individual atoms are refined, either as isotropic values, or even as an anisotropic motion. This anisotropy of vibration is represented by an ellipsoid on each atom.

4.4 Finale

You should be aware that some fundamental developments are currently taking place in x-ray and neutron diffraction with the building recently of exciting new sources of radiation dedicated to diffraction studies. In the x-ray field, there is now considerable interest in synchrotron radiation produced by electrons being accelerated around a circle. Accelerating charged particles produce electromagnetic radiation (Larmor radiation), and when the electrons are made to travel at near relativistic speeds this radiation streams out ahead of the electron in a finely-collimated beam. In addition to being highly collimated, the beam is pulsed, plane-polarised, extremely intense and completely white, spanning wavelengths from the hard x-ray end to the visible region. In Britain, for instance, there is a synchrotron source at Daresbury, built especially for the exploitation of its radiation. In many ways, the synchrotron is a perfect source of x-rays. We can expect to see many important developments in this field in the years to come.

A similar advance is now being made in neutron diffraction, with the construction of neutron spallation sources, such as the one at the Rutherford–Appleton Laboratory in England. In this source, protons that have been accelerated in a synchrotron are fired at a lump of U^{238}. This produces an intense, white, divergent and pulsed flux of neutrons. Recent experiments with this source, which at the time of writing has just begun to be used for serious research purposes, look highly interesting.

Bibliography

Easy further reading:

Bragg L 1975 *The Development of X-ray Analysis* (London: Bell)

Dent Glasser L S 1977 *Crystallography and its Applications* (New York: Van Nostrand)

Glusker J P and Trueblood K N 1985 *Crystal Structure Analysis* (Oxford: OUP)

Harburn G, Taylor C A and Welberry T R 1975 *An Atlas of Optical Transforms* (London: Bell)

Ladd M F C 1979 *Structure and Bonding in Solid State Chemistry* (Chichester: Ellis Horwood)

McKie D and McKie C 1986 *Essentials of Crystallography* (Oxford: Blackwell)

Whittaker E J W 1981 *Crystallography: An Introduction for Earth Science (and Other Solid State) Students* (Oxford: Pergamon)

Windle A 1977 *A First Course in Crystallography* (London: Bell)

and if you want more . . .

Barrett C S and Massalski T B 1966 *Structure of Metals* (New York: McGraw-Hill)

Burns G and Glazer A M 1978 *Space Groups for Solid State Scientists* (New York: Academic)

Cullity B D 1967 *Elements of X-ray Diffraction* (Reading, MA: Addison-Wesley)

International Tables for Crystallography 1983 Vol. A, by International Union of Crystallography (Dordrecht: Reidel)

Weinreich G 1965 *Solids: Elementary Theory for Advanced Students* (New York: Wiley)

Woolfson M M 1970 *X-ray Crystallography* (Cambridge:CUP)